Raw materials for glass melting

By
Bo Simmingsköld

Revision, editing and additional translation by
K. H. Teisen & R. D. Wright

Society of Glass Technology
Sheffield, UK

Raw materials for glass melting by Bo Simmingsköld

First published in 1963 as *Råvaror för glassmältning* by Glasforskningsinstitutet, Växjö, Sweden.

Revised by
Bo Simmingsköld, Professor, Tekn. Dr, Hon FSGT

Additional revision, editing and translation by
K. H. Teisen, CEng, FInstE, FSGT
Teisen Furnaces Ltd, Birmingham, UK
R. D. Wright, BSc, FSGT
Consultant in Glass Technology, Eastleigh, Hampshire, UK

The objects of the Society of Glass Technology are to encourage and advance the study of the history, art, science, design, manufacture, after treatment, distribution and end use of glass of any and every kind. These aims are furthered by meetings, publications, the maintenance of a library and the promotion of association with other interested persons and organisations.

Society of Glass Technology
9 Churchill Way, Chapeltown
Sheffield S35 2PY, UK
Tel   +44(0)114 263 4455
Email   info@sgt.org
Web   http://www.sgt.org

The Society of Glass Technology is a registered charity no. 237438.

ISBN 0-900682-93-3

Cover image adapted from AGC Glass Europe © photograph.

RAW MATERIALS FOR GLASS MELTING

CONTENTS

FOREWORD

This handbook will serve as a practical guide on raw materials used in glass melting to those concerned with batch handling, glass melting, glass compositions, purchase of raw materials, etc.

The raw materials are presented in alphabetical order of their chemical names (e.g. soda ash is found under 'sodium carbonate'. Silica sand under 'silicon dioxide' and so on). The alphabetical entry makes it easy to find the actual material starting with its chemical name.

The raw material descriptions contain the following information:

1. Chemical name.
2. Chemical formula.
3. Chemical name in French and German.
4. General description and in actual cases properties of importance for handling, storage etc. (e.g. ability to take up water from the atmosphere, toxity and so on).
5. Quality requirements. These recommendations are based on the author's practical experience. There are some differences from the national standard quality recommendations of various nations which can be obtained from the standards institutions.
6. Application (very special applications have not been included here).
7. Conversion factors for calculation of batch or glass compositions or change to an alternative raw material for the actual glass component. These factors are calculated from the chemical composition of the various materials. In practice the composition differs considerably from the theoretical. Thus the actual chemical analysis should be the basis for batch calculations and judgement of pricing. The analysis ought to express the composition as a percentage of glass forming oxides. Sometimes the analysis obtained from a supplier gives the percentage as elements (%Fe, Ca, Na, etc.) and has to be converted into oxides (Fe_2O_3, CaO, Na_2O, etc.).
8. Notes. Practical experiences or information concerning the use of the actual material.

Bo Simmingsköld Professor, Tekn. Dr, Hon FSGT

MATERIALS INDEX

MATERIALS INDEX

MATERIALS INDEX

RAW MATERIAL DESCRIPTION

ALUMINIUM HYDROXIDE, ALUMINA HYDRATE

$Al_2O_3.3H_2O$
French: Hydroxyde d'aluminium (d'alumine)
German: Aluminiumhydroxyd, Tonerdehydrat

Light white powder. Bulk density ~1.30.

QUALITY REQUIREMENTS
64–65% Al_2O_3.

APPLICATION
One of the main raw materials for introducing alumina (Al_2O_3) in glass. Unlike other sources for Al_2O_3, such as feldspar, nepheline and kaolin, alumina hydrate introduces no glass forming oxide other than Al_2O_3 and can be obtained with comparatively constant composition and low iron content (0.01% Fe_2O_3).

Note: Alumina (Al_2O_3) dissolves only very slowly in the glass melt and is therefore usually not used as a raw material for glass.

CONVERSION FACTORS

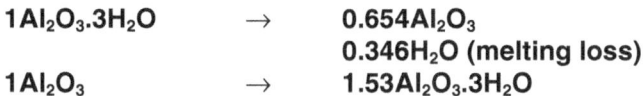

$1Al_2O_3.3H_2O$	$\rightarrow$	$0.654Al_2O_3$
		$0.346H_2O$ (melting loss)
$1Al_2O_3$	$\rightarrow$	$1.53Al_2O_3.3H_2O$

ALUMINIUM NITRATE

$Al(NO_3)_3.9H_2O$
French: Nitrate d'aluminium (d'alumine)
German: Aluminiumnitrat, Tonerdesalpeter

White salt, hygroscopic.

APPLICATION
Refining agent in alkali free glasses.
Raw material for some special glasses.

CONVERSION FACTORS
$1Al(NO_3)_3.9H_2O$ $\rightarrow$ $0.136Al_2O_3$
                                    0.863 (melting loss)

ALUMINIUM SILICATE

$Al_2O_3.2SiO_2.2H_2O$

Kaolin, china clay

French: Terre à porcelaine, Terre de chine
German: Kaolin

Mineral. White powder. Composition varies with deposit.
Actual analysis should be obtained from supplier.

APPLICATION
Raw material for Al_2O_3 (alumina). Used for fluoride opal glasses and E-glass.

CONVERSION FACTORS
$1Al_2O_3.2SiO_2.2H_2O$ $\rightarrow$ $0.395Al_2O_3$
                                            $0.465SiO_2$
                                            $0.139H_2O$ (melting loss)

ANTIMONY

Sb

'Antimony regulus'
French: Antimoine
German: Antimon

Lustrous black-blue metallic powder.

QUALITY REQUIREMENTS
'Pure' antimony should contain ca. 99.5% Sb and max. 0.6% impurities of which Cu and Fe max. 0.01%.
Common grain size 20 mesh US.

APPLICATION
Weak reducing agent, e.g. in melting of some coloured glasses. Sometimes used as a melting catalyser.

NOTE
Sb_2O_3 would be preferred as a refining agent.

ANTIMONY TRIOXIDE

Sb_2O_3
French: Anhydride antimoneux
German: Antimonoxyd

White powder. Toxic (but much less than As_2O_3). Bulk density ~1.20.

QUALITY REQUIREMENTS
98–99.5% Sb_2O_3. Fe_2O_3 max. 0.015%. Technical grade usually contains 0.5% or more As_2O_3.

APPLICATION
Refining agent in combination with potassium or sodium nitrate. Sb_2O_3 is often used as a substitute for As_2O_3 because of lower toxic effect. Effective as refining agent in relatively low melting glasses. Glasses to be

decorated by silver staining usually contain some Sb_2O_3. Antimony trioxide increases scratch hardness and gives some special optical properties.

NOTE
Sb_2O_3 in combination with nitre has somewhat weaker oxidising effect on Fe in glass than As_2O_3.

ARSENIC TRIOXIDE

As_2O_3

'Arsenic'
French: Anhydride arsènieux
German: Arsentrioxyd, 'Arsenikmehl'

White powder. Very toxic. Special regulations for handling and storage.

QUALITY REQUIREMENTS
$As_2O_3 > 99.5\%$, $Fe_2O_3 > 0.02\%$

APPLICATION
Refining and 'chemical' decolourising agent in combination with potassium or sodium nitrate.

NOTE
15–40% of the As_2O_3 introduced into the batch evaporates from the melting batch when melting in open pots (for Sb_2O_3 the loss is in the order of 10%).

CONVERSION FACTORS
$$1As_2O_3 \quad \rightarrow \quad 1.16As_2O_5$$

AURIC CHLORIDE

$AuCl_3.4H_2O$
French: **Triclorure d'or**
German: **Goldchlorid**

Brown-red crystals. Soluble in water. Contains theoretically 58.1% Au.

APPLICATION
Colouring agent in gold ruby glasses. A weak hydrochloric acid water solution is usually mixed in the batch.

NOTE
Sometimes another gold compound $HAuCl_4.4H_2O$(47.7% Au) is used. 'Cassius Purple' which is a stannic acid compound containing various amounts of gold (20–75% Au) was also used in olden times.

BARIUM CARBONATE

$BaCO_3$
French: **Carbonate de baryte**
German: **Bariumkarbonat**

White powder or micropellets. Insoluble in water. Toxic. Bulk density ~0.6.

NOTE
Barium carbonate has a tendency to agglomerate during batch mixing. The problem can be overcome by using micropelletised material.

QUALITY REQUIREMENTS
Should contain 98–99% $BaCO_3$. Iron content max. 0.015% Fe_2O_3. The material often contains up to 0.5% barium sulphide BaS and/or some barium sulphate $BaSO_4$. 'Sulphur free' qualities (~0.04% BaS) are available on the market.
Witherite is a mineral which mainly contains $BaCO_3$. It is used as a raw material in production of pure barium carbonate.

APPLICATION
Main raw material for introducing barium oxide, BaO, in glass.

CONVERSION FACTORS

$1BaCO_3$ $\rightarrow$ $0.777BaO$
$0.223CO_2$ (melting loss)
$1BaO$ $\rightarrow$ $1.29BaCO_3$

BARIUM FLUORIDE

BaF_2
French: Flourure de barium
German: Bariumfluorid

White powder.

APPLICATION
Fluxing and refining agent sometimes used in tank melting.

BARIUM METAPHOSPHATE

$Ba(PO_3)_2$
French: Métaphosphate de barium
German: Bariummetaphosphat

APPLICATION
Raw material for the introduction of P_2O_5. Used for some phosphate opal glasses.

CONVERSION FACTORS

$1Ba(PO_3)_2$ $\rightarrow$ $0.52BaO$
$0.48P_2O_5$

BARIUM NITRATE

$Ba(NO_3)_2$
French: Salpêtre de baryte
German: Bariumnitrat, Barytsalpeter

Crystalline white salt. Toxic. Barium nitrate tends to harden under its own weight during storage.

QUALITY REQUIREMENTS
Min. 99% $Ba(NO_3)_2$, the rest is normally sodium nitrate and water. $Fe_2O_3 < 0.002\%$.

APPLICATION
Oxidising and refining agent in combination with arsenic trioxide or antimony trioxide, especially for alkali free glasses or glasses with low alkali content.

CONVERSION FACTORS
$1Ba(NO_3)_2$	$\rightarrow$	$0.583BaO$
		0.415 (melting loss)
$1BaO$	$\rightarrow$	$1.70Ba(NO_3)_2$

BARIUM SELENITE

$BaSeO_3$
French: Sélenite de baryum
German: Bariumselenit

White salt, insoluble in water. Toxic.

QUALITY REQUIREMENTS
Min. 28% Se.

APPLICATIONS
Decolourising. Colouring agent (selenium rose, selenium ruby in combination with cadmium sulphide).

CONVERSION FACTORS
$1BaSeO_3$ $\rightarrow$ 0.299Se
                              0.580BaO

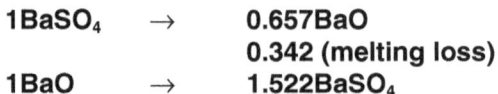

BARIUM SULPHATE

$BaSO_4$
French: Sulfate de baryum
German: Bariumsulfat

White powder. Bulk density ~1.78. Insoluble in water.

QUALITY REQUIREMENTS
97–98% $BaSO_4$, $Fe_2O_3 < 0.025\%$.

APPLICATION
Refining agent.

CONVERSION FACTORS
$1BaSO_4$ $\rightarrow$ 0.657BaO
                              0.342 (melting loss)
$1BaO$ $\rightarrow$ 1.522$BaSO_4$

BERYLLIUM CARBONATE

$BeCO_3$
French: Carbonate de beryllium
German: Berylliumkarbonat

White powder.

APPLICATION
Raw material for the introduction of beryllium oxide (BeO) in glass. Beryllium oxide increases surface hardness in glass and e.g. is sometimes used in jewellery glass articles.

CONVERSION FACTORS
$1BeCO_3$ $\rightarrow$ $0.363BeO$
$0.637CO_2$ (melting loss)
$1BeO$ $\rightarrow$ $2.75BeCO_3$

BISMUTH OXIDE

Bi_2O_3
French: Oxyde de bismuth
German: Wismuttrioxyd

Lemon-yellow powder.

APPLICATION
Introduction of Bi_2O_3 in glass (e.g. some optical glasses with high refractive index). Small amounts are sometimes used in copper-ruby batches (to control the rate of colour development).

BORIC ACID

H_3BO_3
French: Acide borique
German: Borsäure

White powder. Flaky crystals. Bulk density ~1.0.

QUALITY REQUIREMENTS
99.5–100% H_3BO_3, Fe_2O_3<0.001%.

APPLICATION
Introduction of B_2O_3 in glass. Used for high B_2O_3 low R_2O glasses e.g. Pyrex, E-glass.

CONVERSION FACTORS
$1H_3BO_3$ → $0.563B_2O_3$
                    $0.437H_2O$ (melting loss)
$1B_2O_3$ → $1.79H_3BO_3$

NOTE
Some B_2O_3 normally volatilises from the melt. The loss increases with the water content in the batch and the furnace atmosphere.

CADMIUM SULPHIDE

CdS
French: Sulfure de cadmium
German: Cadmiumsulfid

Yellow or orange-yellow powder. Toxic.

QUALITY REQUIREMENTS
The material should contain min. 98% CdS and the rest moisture. The sulphide should be soluble in 20% hydrochloric acid (absence of impurities such as free sulphur, kaolin, barium sulphate etc.). Normal commercial quality contains max. 0.02% Fe.

APPLICATION
Colouring agent for yellow glasses and in combination with selenium for red selenium ruby glasses. Up to about 50% of the sulphide in the batch could evaporate during melting.

CALCIUM BORATE, COLEMANITE

$Ca_2B_6O_{11}.5H_2O$
French: Borate de calcium
German: Kalziumborat

The natural mineral is used in the glass industry. It is also a raw material for the production of other pure boron compounds such as boric acid. Colemanite, borocalcite, contains:
43–49% B_2O_3, 25–28% CaO
0.3–2% MgO, 0.05–5% SiO_2, 0.3–0.5% Na_2O+K_2O
0.4–0.8% Al_2O_3, 0.2–0.7% Fe_2O_3.
Actual analysis should be obtained from supplier.

APPLICATION
Raw material for B_2O_3 in glass. It is often used in the fibre glass production ('E-glass').
CONVERSION FACTORS
$1Ca_3B_6O_{11}.5H_2O$ $\rightarrow$ B_2O_3
CaO
H_2O (melting loss)

CALCIUM CARBONATE, LIME

$CaCO_3$
French: Carbonate de calcium, chaux
German: Kalziumkarbonat, Kalk, Kreide

Calcium carbonate is normally used in the form of ground limestone, chalk or marble. The composition varies from different deposits.
Actual analysis should be obtained from supplier.

QUALITY REQUIREMENTS
Composition
'Pure' limestone should contain min. 55.0% CaO (=98.1% $CaCO_3$) and have the specification given for Quality 1 (see below).
'Ordinary limestone' should contain min. 47.7% CaO (equal to 85% $CaCO_3$). The rest should mainly be MgO, SiO_2 and Al_2O_3.
'Dolomitic limestone' should contain about 15% MgO and min. 85% $CaCO_3+MgCO_3$.
Grain distribution
The main part of the material normally lies between 1–0.005 mm.

	Bottles (%)	Flat glass (%)	Pot melting (%)
>1 mm	max. 0.5		
>0.5 mm		max. 0.5	max. 0.5
<0.1 mm	max. 10	max. 25	max. 30

Analytical requirements based on the content of iron oxide Fe_2O_3

Quality 1 Fe_2O_3 max 0.04%, max 0.0004% Cr_2O_3 (recommended for high quality 'colourless' glasses)
Quality 2 Fe_2O_3 max 0.12% (ordinary 'colourless' glasses)
Quality 3 Fe_2O_3 max 0.3%
Quality 4 Fe_2O_3>0.3%

The material must be free from foreign particles. Chalk sometimes contains particles of flint which may cause stones in the glass.
Lime for the production of colourless glass must not contain significant amounts of other colouring metals e.g. chromium (which may contaminate the material when steel balls are used in the grinding of the mineral). Usual impurities are SiO_2, Al_2O_3, MnO, SO_3 and organic matter. Actual analysis should be obtained from supplier.
In Quality 1 and 2 matter insoluble in hydrochloric acid should not exceed 1%. Moisture content ought not to exceed 2%.
The composition must not vary much within one delivery: ±0.5% CaO and ±0.02% Fe_2O_3 (in Quality 1–3).

APPLICATION
Main raw material for the introduction of CaO in glass.

NOTE
Sometimes, calcium oxide, 'quick lime', CaO is used. For instance, this is the case for electric melting in 'cold top furnaces' of fibre glass.

CONVERSION FACTORS
$$1CaCO_3 \quad \rightarrow \quad 0.560CaO$$
$$0.440CO_2 \text{ (melting loss)}$$
$$1CaO \quad \rightarrow \quad 1.785CaCO_3$$

CALCIUM FLUORIDE, FLUORSPAR

CaF_2
French: Fluorure de calcium
German: Kalziumfluorid, Flusspat

QUALITY REQUIREMENT
Normally the pulverised mineral is used (60–90% CaF_2). The material contains various amounts of $CaCO_3$, SiO_2 and Al_2O_3. The iron content also varies: 0.1–0.4% Fe_2O_3. Sometimes fluorspar contains some $BaSO_4$ which could influence refining.

APPLICATION
Opalising agent. Small amounts of fluorspar are used as fluxing agents and to improve refining.

NOTE
Some fluorine always evaporates during the melting as SiF_6 or alkali fluorides. This can result in serious pollution problems.

CONVERSION FACTORS
$$1CaF_2 \quad \rightarrow \quad 0.486F_2$$
$$0.718CaO$$

CALCIUM MAGNESIUM ALUMINIUM SILICATE, 'SLAG'
French: Laitier
German: Schlacke

'Slag', usually from blast furnaces, is supplied as granular powder with various compositions. (Actual analysis should be obtained from supplier.) It normally contains calcium aluminium silicate ($Ca_2Al_2SiO_7$), magnesium calcium silicate ($Ca_2MgSi_2O_7$), calcium silicate ($CaSiO_3$) and some sulphites (1–2%). The CaO content is around 41–44%.

QUALITY REQUIREMENTS

To ensure constant chemical composition the supplier has to blend large quantities of screened slag. (Actual analysis and limits of variations of the composition should be obtained from supplier.) Example of a 'bene-ficiated' slag product ('Calumite'):
36% SiO_2, 13% Al_2O_3, 8% MgO, 40.5% CaO, 0.4% MnO, 0.85% S, 0.3% SO_3, 0.24% Fe_2O_3, 0.3% Na_2O, 0.4% K_2O, 0.5% TiO_2.

APPLICATION

To replace other CaO and Al_2O_3 raw materials. The slag is a 'pre-reacted' material and contains some sulphides. Because of this the slag acts as a fluxing agent and is said to reduce 'the time for batch free melt' and this increases the possible output per unit melting area in a tank.

CALCIUM METASILICATE

$CaSiO_3$
French: Mètasilicate de calcium
German: Kalziummetasilikat

White powder. In glass making the natural mineral β-wollastonite is usually used.
Example of wollastonite (Finnish):
51.80% SiO_2, 44.50% CaO, 0.44% Al_2O_3, 0.22% Fe_2O_3,<0.05% TiO_2, <0.01% MnO, 0.56% MgO, 0.10% Na_2O, 0.10% K_2O, loss on ignition, 2.20, bulk density ~1.3.

APPLICATION

Raw material for introduction of CaO. Main use in electric melting when a minimum of gases from batch reactions is wanted.

CONVERSION FACTORS

$$1CaSiO_3 \quad \rightarrow \quad 0.483CaO$$
$$0.517SiO_2$$

CARBON

C
French: Carbone
German: Kohlenstoff

APPLICATION
Pulverised stonecoal, charcoal or graphite are used as reducing agent in 'carbon amber' glasses. (The colouring agent is actually iron sulphide). The composition (content of sulphur) and the grain size influences the colouring effect.

CERIUM OXIDE

CeO_2
French: Oxyde cérique
German: Ceriumoxyd

Pure cerium oxide is a yellow-white powder. Technical qualities are usually brown and contain various amounts of Nd_2O_3, Pr_2O_3, La_2O_3.

APPLICATIONS
Refining and decolorising agent. Yellow glasses in combination with titanium dioxide ('cerium-titanium yellow'). Cerium oxide increases ultraviolet absorption in glass. It is used to avoid discoloration (darkening) from nuclear radiation and x-rays in glass. Photochromic glasses contain some CeO_2.

NOTE
Cerium oxide in combination with arsenic trioxide (As_2O_3) makes glass very sensitive to solarisation.

CHROMIC OXIDE

Cr_2O_3
French: Oxyde de chrome
German: Chromoxyd

Green powder. Cr_2O_3 dissolves very slowly in the melt which is why potassium chromate is more often used.

APPLICATION
Green colour in glass. Soda-lime-glasses are coloured grass-green, lead glasses green with a yellow tint.
Chromic oxide is a strong colouring agent. 0.05% Cr_2O_3 in a glass gives a distinct grass-green colour.

NOTE
Chromite (chrome ironstone) is used as a colouring agent where iron is acceptable in the batch e.g. in olive or blue-green bottles. Typical analysis 42–44% Cr_2O_3, 4–5% SiO_2, 25% FeO, 8–10% MgO, 16–18% Al_2O_3.

COBALTOUS OXIDE

CoO
French: Oxyde cobalteux
German: Kobaltoxyd

The mostly used compound is 'black cobalt oxide' which is a mixture of CoO and Co_2O_3 containing 70–71% Co. 'Grey cobalt oxide' usually contains 75–76% Co. Actual analysis should be obtained from supplier.

QUALITY REQUIREMENTS
Min. 70% Co. Common impurities are Ni (~0.25%), Fe (~0.05%) and Mn (~0.2%).

APPLICATION
1. Colouring agent for blue glasses. Soda-lime-glasses are blue with a red-violet tint, potassium glasses with a blue-violet tint. Borosilicate glasses become a red-violet colour. Cobalt oxide is a very strong colouring agent.
2. Decolourising agent (0.1–0.2 g/100 kg of silica sand) in combination

with nickel oxide, selenium or manganese dioxide. To get a homogeneous distribution of CoO in the batch, mixtures of lime or feldspar containing some % of CoO are used.

NOTE
'Powder blue' (German 'Schmalte') is a pulverised cobalt-potassium glass with 3–8% CoO content (actual analysis should be obtained from supplier). This material is sometimes used instead of cobalt oxide mixtures.

COPPER OXIDE

CuO

Black copper oxide
French: **Oxyde cuivrique anhydré**
German: **Kupferoxyd**

Black powder (<200 mesh).

QUALITY REQUIREMENTS
The CuO content should be 95–99%. Metallic Cu must not be present. Extremely fine-grained qualities should be avoided as they sometimes cause refining trouble in glasses containing significant amounts of copper oxide. Typical analysis: 99% CuO, 0.5% Cu_2O, 0.23% Fe, 0.05% Zn.

APPLICATION
Colouring agent for blue glasses. With increasing CuO concentration the colour moves from pure sky blue with some green tint. Lead glass becomes blue-green, soda-lime glasses become blue.

NOTE
For production of glasses with high CuO content (>1.5% CuO) it is recommended that some of the copper be introduced as copper sulphate (to avoid incomplete solution in the melt and black stones in the glass).

COPPER SULPHATE

$CuSO_4.5H_2O$
French: Sulfate de cuivre
German: Kupfersulfat

Blue crystalline powder soluble in water.

QUALITY REQUIREMENTS
98–99% $CuSO_4+5H_2O$ and max. 1% Fe_2O_3.

APPLICATION
Blue colour in glass (see copper oxide).

NOTE
High content of copper sulphate in the batch can sometimes cause refining troubles (SO_3). The waterfree salt $CuSO_4$ has also been used.

CONVERSION FACTORS
$1CuSO_4.5H_2O$	$\rightarrow$	$0.319CuO$
		$0.321SO_3$ (melting loss)
		$0.361H_2O$ (melting loss)
$1CuO$	$\rightarrow$	$3.138CuSO_4.5H_2O$

CUPROUS OXIDE

Cu_2O

Red copper oxide
French: Oxyde cuivreux
German: Kupferoxydul

Red powder.

APPLICATION
Colouring agent in copper-ruby glass. To get the ruby the batch should contain reducing agents and the melting condition should be reducing.

DICALCIUM ORTHOPHOSPHATE

$CaHPO_4.2H_2O$
French: Phosphate bicalcique
German: Dikalziumphosphat

White powder.

APPLICATION
Opalising agent in phosphate opal glass.

CONVERSION FACTORS
$1CaHPO_4.2H_2O$ $\rightarrow$ $0.326CaO$
$0.413P_2O_5$
$0.261H_2O$ as melting loss

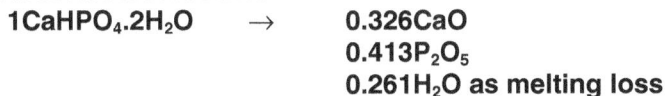

DIDYMIUM OXIDE
French: Oxyde de didyme
German: Didymoxyd

Brown powder composed of mixtures of neodymium oxide, praseodymium oxide and some other rare earth oxides e.g. lanthanum and samarium oxide. Actual analysis should be obtained from supplier.
Example: 52% Nd_2O_3, 14% Pr_6O_{11}, 19% La_2O_3, 8% Sm_2O_3, 0.2% Ce_2O_3, the rest 6%.

APPLICATIONS
Colouring agent, violet-blue, blue-green depending on composition.
Shows dichroism.
Not suitable for decolourisation because of the praseodymium content.

DISODIUM TETRABORATE

$Na_2B_4O_7$
French: Borate de sodium anhydre, borax anhydre
German: Natriumborat wasserfrei

White powder, rather hygroscopic. Should be stored dry.
'Dehybor' is the commercial name for this material which is also known as anhydrous borax and containing approx 68.5% B_2O_3. See also disodium tetraborate decahydrate.

APPLICATION
Main raw material for B_2O_3. (See also disodium tetraborate decahydrate.)
Small amounts of B_2O_3 are used to speed up melting and refining.

NOTE
There is another quality of borax on the market called 'calcined borax' containing 78% $Na_2B_4O_7$ equal to 54% B_2O_3 and 24% Na_2O. It is always necessary to ask for analysis from supplier.

CONVERSION FACTORS
$1Na_2B_4O_7$	$\rightarrow$	$0.692B_2O_3$
		$0.308Na_2O$
$1B_2O_3$	$\rightarrow$	$1.45Na_2B_4O_7$
$1Na_2B_4O_7.10H_2O$	$\rightarrow$	$0.53Na_2B_4O_7$

DISODIUM TETRABORATE DECAHYDRATE

$Na_2B_4O_7.10H_2O$
French: Borax decahydraté
German: Kristallborax

White crystalline salt (prismatic). Delivered as fine granulated material or powder.

APPLICATION
Main raw material for B_2O_3. (See also disodiumtetra borate)

NOTE

There is also an octahedral form of borax containing only $5H_2O$ (47.8% B_2O_3, 21.3% Na_2O). The chemical name is disodium teraborate pentahydrate and the commercial name is Neobor. This is rarely used in the glass industry. In many cases natural B_2O_3 containing minerals are used. A common such 'natural' borax is 'Rasorite' (typical analysis: 47.6% B_2O_3, 21.5% Na_2O, 0.4% Al_2O_3, 1.4% SiO_2, 0.1% Fe_2O_3, 27.9% H_2O). Another is 'Colemanite' $Ca_2B_6O_{16}$ (typical analysis: 50–45% B_2O_3, 28–25% CaO, 0.2–2.5% MgO, 0.5–5.5% SiO_2, 0.4–0.6% R_2O, 0.4–0.6% Al_2O_3).

CONVERSION FACTORS

$1Na_2B_4O_7.10H_2O \rightarrow$ $\quad$ $0.365B_2O_3$
$\qquad\qquad\qquad\qquad\qquad\quad$ $0.163Na_2O$
$\qquad\qquad\qquad\qquad\qquad\quad$ $0.473H_2O$ **(melting loss)**
$1B_2O_3 \qquad\qquad\quad \rightarrow$ $\quad$ $2.73Na_2B_4O_7.10H_2O$

ERBIUM OXIDE

Er_2O_3
French: **Oxyde de erbium**
German: **Erbiumoxyd**

Rose-red powder.

APPLICATION

Rose-red colour in glass. Ideal physical decolourising agent. Because of weak colouring power and high price only used for expensive colourless glasses, normally in combination with neodymium oxide (e.g. 0.003% Er_2O_3+0.008% Nd_2O_3 in glasses containing max. 0.015% Fe).

FERROSOFERRIC OXIDE

Fe_3O_4
French: Oxyde ferroso-ferrique
German: Eisenhammerschlagg

Black powder.

APPLICATION
Gives blue green colour in glass. Has sometimes been used as reducing agent (e.g. in copper-ruby batches)

IRON OXIDE

Fe_2O_3
French: Oxyde de fer, 'Potée rouge'
German: Eisenoxyd, Eisenoxyd rot

Brown or red powder. In commercial qualities the iron content varies and they may also contain carbonate and water (97–25% Fe_2O_3). Both natural and synthetic iron oxide is used. The latter holds, as a rule, about 97% Fe_2O_3.

APPLICATION
Colouring agent. Reducing conditions give blue green and oxidising yellow-green colour. In combination with manganese or sulphides it gives a brown colour. For so called carbon-amber coloured glasses (bottles) carbon and sodium sulphate are added to the batch. The colouring compound in these glasses is iron sulphide. The small amounts of iron from impurities in the raw material are normally enough to produce the colour.

NOTE
In the glass both Fe_2O_3 and FeO exist together. The equilbrium depends on the redox conditions in the melt.

IRON SULPHIDE (PYRITES)

FeS_2
French: Pyrite
German: Schwefelkies

Pulverised yellow mineral containing 63.6% S, 46.7% Fe.

APPLICATION
Colouring agent for amber or brown glasses.

LEAD OXIDE

PbO

Litharge
French: Oxyde de plomb
German: Bleioxyd

Yellow powder. Toxic. Bulk density ~2.4. Sometimes used in the form of 'flakes' or micropellets to reduce dusting.

QUALITY REQUIREMENTS
Fe_2O_3<0.003% for colourless glasses. Must not contain metallic lead.

APPLICATION
Main raw material for the introduction of PbO in glass. Litharge has the disadvantage, compared with red lead, of lower oxygen content and is therefore more sensitive to reduction to metallic lead in the melt.

LEAD SILICATE

xPbO.ySiO$_2$
French: Silicate de plomb
German: Bleisilikat

Yellow-white glassy grains.

	PbO (%)	SiO$_2$ (%)	Al$_2$O$_3$ (%)	Fe$_2$O$_3$ (%)
Lead monosilicate	84.5	15.5	0.25	0.01
Lead disilicate	65	32.2	2.8	0.02

The most commonly used compound is the monosilicate PbO.SiO$_2$. Lead silicate is normally delivered as a rather coarse ground product (~2–3 mm dia.) containing some percent of water. Care must be taken to avoid mixing problems. The main advantage with this product is that dusting can be avoided. Fine dry powder of lead silicate has a high tendency towards dusting.

CONVERSION FACTORS

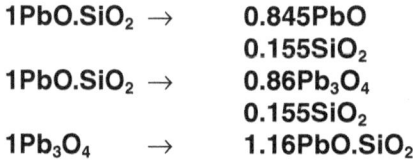

1PbO.SiO$_2$ $\rightarrow$ 0.845PbO
                           0.155SiO$_2$
1PbO.SiO$_2$ $\rightarrow$ 0.86Pb$_3$O$_4$
                           0.155SiO$_2$
1Pb$_3$O$_4$ $\rightarrow$ 1.16PbO.SiO$_2$

These factors are based on the analysis given above and not the chemical formula.

LEAD II PLUMBATE

Pb_3O_4

Red Lead, Minium
French: Minium de plombe
German: Bleimennige, Mennige

Red-red orange fine heavy powder. Bulk density ~3.15. Dusty in spite of high density. Toxic.

QUALITY REQUIREMENTS
For colourless lead crystal glasses:
Fe_2O_3<0.003%
PbO_2>28%
Soluble in hydrochloric acid.

APPLICATION
Main raw material for the introduction of lead oxide PbO in glass.

NOTE
To reduce the incidence of dust, red lead is often moistened with water (3–5%) or light oil (<1%). Red lead in the form of micro-pellets are also available.
In order to reduce the environmental risks, pelletising of batches rich in lead oxide has been introduced.

CONVERSION FACTORS
$1Pb_3O_4$	$\rightarrow$	$0.977PbO$
$1PbO$	$\rightarrow$	$1.02Pb_3O_4$

LITHIUM ALUMINIUM SILICATE
French: **Lithium aluminum silicate**
German: **Lithium Aluminium Silikat**

Some lithium aluminium silicate minerals are used as glass raw materials
for Li_2O lithium oxide. The most common are:
1. Spodumene $Li_2O.Al_2O_3.4SiO_2$ (~8.4% Li_2O)
2. Lepidolite $(Li, K)_2 Al_2, (F, OH)_2.Si_3O_9$ (~6.4% Li_2O)
3. Petalite $Li_2O.Al_2O_3, 8SiO_2$ (~4% Li_2O)
The minerals are also raw materials for the production of lithium carbonate.

APPLICATION
Introduction of Li_2O (and Al_2O_3) in glass.

LITHIUM CARBONATE

Li_2CO_3
French: **Carbonate de lithium**
German: **Lithiumkarbonat**

White powder. Common quality 98%Li_2CO_3.

APPLICATION
Main raw material for Li_2O in glass. Lithium oxide is a strong fluxing agent.
Reduces viscosity of the melt.
Lithium oxide is also a component in many (low expansion) glass ceramics
($Li_2O–Al_2O_3–SiO_2$ and $Li_2O–ZnO–P_2O_5–SiO_2$ systems).

CONVERSION FACTORS
$1Li_2CO_3$ $\rightarrow$ $0.404Li_2O$
                  $0.596CO_2$ (melting loss)
$1Li_2O$ $\rightarrow$ $2.473Li_2CO_3$

MAGNESIUM CALCIUM CARBONATE

$MgCO_3.CaCO_3$

Dolomite
French: Dolomie
German: Dolomit

The granular mineral is used as a raw material.

QUALITY REQUIREMENTS
See under calcium carbonate.

APPLICATION
Main raw material for MgO in glass as long as the CaO content of the glass in question allows its use.

CONVERSION FACTORS

$1MgCO_3.CaCO_3$	$\rightarrow$	$0.217MgO$
		$0.304CaO$
		$0.479CO_2$ (melting loss)
$1MgO$	$\rightarrow$	$4.61MgCO_3.CaCO_3$

MAGNESIUM CARBONATES

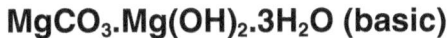

$MgCO_3$ (neutral)

$MgCO_3.Mg(OH)_2.3H_2O$ (basic)
French: Carbonate de magnesium
German: Magnesiumcarbonat

White light powders. The MgO content varies much in different materials. Actual analysis should be obtained from supplier. Common MgO content 40–48% MgO.

APPLICATION
Main raw materials for MgO in calcium free or low calcium glasses.

NOTE
'Magnesia alba' is a very light carbonate (~41% MgO). 'Heavy' magnesium carbonate is a more dense quality (~46% MgO).
The natural mineral 'magnesite' ($MgCO_3$) has sometimes been used. Magnesium oxide (MgO) is not often used as a glass raw material.

CONVERSION FACTORS

$1MgCO_3$ $\rightarrow$ **0.478MgO**
                            0.522CO$_2$ (melting loss)
1 MgO $\rightarrow$ **2.092MgCO$_3$**

MAGNESIUM SILICATE

$3MgO.4SiO_2.H_2O$

Talc, soapstone
French: **Talc, Silicate de magnesium**
German: **Talk, Speckstein, 'Federweiss'**

White powder. Composition varies with deposit.

APPLICATION
Introduction of MgO in glass. Talc is also used in some kinds of alabaster glasses.

CONVERSION FACTORS

$1(3MgO.4SiO_2.H_2O) \rightarrow$ **0.318MgO**
                                       0.633SiO$_2$

MANGANESE DIOXIDE

MnO_2

Pyrolusite
French: Pyrolusite
German: Braunstein

Black powder. For glass production the mineral pyrolusite (or some other MnO_2 containing minerals) is normally used.

QUALITY REQUIREMENTS
Composition varies. Actual analysis should be obtained from supplier. For decolourising: min. 84% MnO_2, max. 1% Fe_2O_3.

APPLICATION
Colouring agent for violet glasses and together with Fe_2O_3 for brown glasses. In the past it was used as a decolourising agent in combination with nitre and arsenic.

MOLYBDENUM DISULPHIDE

MoS_2
French: Disulfide de molybdine
German: Molybdänsulfid

Blue-black mineral powder.

APPLICATION
Colouring agent. Gives red-orange colour under reducing conditions. Often sodium sulphide is combined with MoS_2 in the batch.

NEODYMIUM OXIDE

Nd_2O_3
French: Oxyde de néodyme
German: Neodymoxyd

Grey powder. Tends to absorb H_2O and CO_2 from the atmosphere.

QUALITY REQUIREMENTS
For decolourising: min. 98% Nd_2O_3, max 1.5% praseodymium oxide, Pr_6O_{11}. If used in arsenic refined glasses the material should not contain cerium oxide to avoid solarisation.

APPLICATION
Decolourising agent. Gives blue-violet colour in glass. The colour shows 'dichroism' which means that the colour looks different at different thicknesses and under different light sources. In combination with selenium a wine-red colour is obtained (neodymium-ruby).

NOTE
Other neodymium compounds such as the hydroxide or oxalate are sometimes used.

NICKEL MONOXIDE

NiO
French: Oxydule de nickel
German: Nickeloxydul

Grey or grey-green powder containing 90–99% NiO. 'Green nickel oxide' is nickel hydroxide $Ni(OH)_2$ (80.5–75% NiO) or some of the carbonates (63–55% NiO). Actual analysis should be obtained from supplier.

APPLICATION
See nickel oxide.

CONVERSION FACTORS
    1NiO   → $1.1Ni_2O_3$

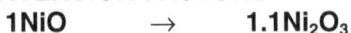

NICKEL SESQUIOXIDE, NICKELIC OXIDE

Ni_2O_3
French: Oxyde de nickel
German: Nickeloxyd

Black powder. The compound $Ni_2O_3 \cdot H_2O$ (90.2% Ni_2O_3) is sometimes used. See also Nickelmon oxide, NiO.

APPLICATION
In potassium and potassium lead glasses violet colour is obtained. Sodium glasses produce a somewhat violet-brown colour. Often used in combination with other colouring oxides (Mn, Cu, Fe) to obtain 'smoky' colours. The main decolourising agent for lead crystal glasses.

NOTE
See also nickel monoxide, NiO.

CONVERSION FACTORS
$1Ni_2O_3$ $\rightarrow$ $0.91NiO$

PHOSPHORIC ACID

H_3PO_4
French: Acide phosphorique
German: Phosphorsäure

White crystals. Very hygroscopic.

APPLICATION
Introduction of P_2O_5 in glass. As phosphoric acid is very hygrosopic it is added to the batch as a concentrated water solution (e.g. 85%). If much phosphoric acid is introduced the batch has to be dried before being charged to the furnace.

NOTE
In alkali-free phosphate glasses aluminium phosphate is sometimes used as a P_2O_5 raw material (58.22% P_2O_5, 41.78% Al_2O_3).

CONVERSION FACTORS

$1H_3PO_4$ $\rightarrow$ $72.44P_2O_5$
$26.56H_2O$ (melting loss)

$1P_2O_5$ $\rightarrow$ $1.38H_3PO_4$

POTASSIUM ALUMINIUM SILICATE

$K_2O.Al_2O_3.6SiO_2$

Feldspar
French: **Feldspath de potassium**
German: **Kalifeldspat**

Feldspar is used as pulverised mineral. The composition varies with deposit. Pure potassium feldspar deposits are rare. Most commercial feldspars are pegmatites, i.e. mixtures of rather coarse crystals of potassium feldspar, sodium feldspar (Albite) and silica. Often the feldspar is separated from silica in the pulverised pegmatite by flotation. Feldspar sometimes also contains mica (MgO, Fe_2O_3). For batch calculation it is always necessary to know the actual chemical composition of the material.

	Scandinavian feldspars composition limits (%)	Finnish feldspar 'FFF' (%)	Pegmatite from Forshammar Sweden (%)
SiO_2	64–68	68	75.9
Al_2O_3	18–19.5	18.5	14.3
Fe_2O_3	0.1–0.4		0.11
CaO	0–0.3	0.04	0.23
MgO	0.1–0.4		0.20
K_2O	9–14	8.2	4.76
Na_2O	1–5	5.2	4.06
Ignition loss	0.1–0.5		0.46

QUALITY REQUIREMENTS
The material should be free of foreign particles, clay, mica, etc.
Grain size:
75% within 0.05–0.3 mm; 0.2%>0.5 mm; 5%<0.05 mm.
From the point of iron content the 'iron number' = $(\%Fe_2O_3)/(\%Al_2O_3)$ is sometimes used.

'Iron number'
Quality I   0.006 ('colourless' glasses)
Quality II   0.012 ('semiwhite' glasses, bottles)
Quality III   >0.012 (coloured glasses).
The loss of ignition ought not to exceed 1% (higher figures may cause refining problems).
The material should be homogeneous in delivered quality:
±0.5% alkali content
±0.25% Al_2O_3

APPLICATION
Main raw material for the introduction of Al_2O_3 in glass.

CONVERSION FACTORS

Forshammar pegmatite	→	**0.142 Al_2O_3**
FFF feldspar	→	**0.185 Al_2O_3**
1Al_2O_3	→	**7.02 'Forshammar' pegmatite**
1Al_2O_3	→	**5.41 'FFF' feldspar**

POTASSIUM ALUMINIUM SULPHATE

$K_2SO_4.Al_2(SO_4)_\backslash.24H_2O$

Alum
French: Alun
German: Kalialun

White powder soluble in water.

APPLICATION
Opalising agent in alabaster glass.

CONVERSION FACTORS

1$K_2SO_4.Al_2(SO_4)_3.24H_2O$	→	**0.1K_2O**
		0.11Al_2O_3
		0.34SO_3 and
		0.46H_2O (melting loss)

POTASSIUM BICARBONATE

$KHCO_3$
French: Bicarbonate de potassium
German: Kaliumbikarbonat

White crystalline powder. Pure bicarbonate (containing no potassium, carbonate is not hygroscopic—in contrast to potash).

QUALITY REQUIREMENTS
Max. 3% K_2CO_3, 0.5% KCl, 0.004% Fe_2O_3, 0.0001% Cr_2O_3 (for colourless glasses).

APPLICATION
Main raw material for potasium oxide, K_2O, in glass.
Storage advantage: pure material is not hygroscopic.

CONVERSION FACTORS
$1KHCO_3$ $\rightarrow$ $0.470K_2O$
                     $0.440CO_2$ (loss)
                     $0.090K_2O$ (loss)
$1K_2O$ $\rightarrow$ $2.126KHCO_3$

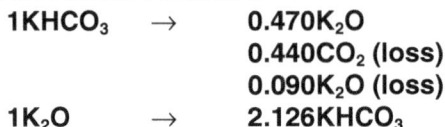

POTASSIUM BICHROMATE

$K_2Cr_2O_7$
French: Bichromate de potassium
German: Kaliumbichromat

Yellow-red crystalline powder. Toxic.

QUALITY REQUIREMENTS
99.5–99.8% $K_2Cr_2O_7$

APPLICATION
Colouring agent for green glasses. (See also Cr_2O_3). Easier to get into solution in the melt than chromium oxide.

NOTE

Often potassium chromate, K_2CrO_4, is used instead of the bichromate.

CONVERSION FACTORS

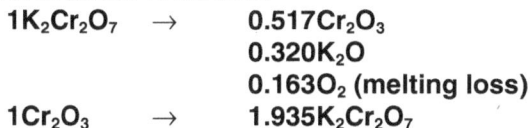

$1K_2Cr_2O_7$ $\rightarrow$ $0.517Cr_2O_3$
$0.320K_2O$
$0.163O_2$ **(melting loss)**
$1Cr_2O_3$ $\rightarrow$ $1.935K_2Cr_2O_7$

POTASSIUM CARBONATE

K_2CO_3

Potash dehydrated
French: Carbonate de potassium
German: Kaliumcarbonat, Pottasche

White crystalline powder. Very hygroscopic. Tends to harden even in closed containers. Bulk density approx 1.1.

QUALITY REQUIREMENTS

96–98% K_2CO_3. Max 0.0005% Fe_2O_3. Max 0.0001% Cr_2O_3 (colourless glasses).
Other components see under: Potassium carbonate. Hydrated.

APPLICATION

Main raw material for K_2O in glass.

CONVERSION FACTORS

$1K_2CO_3$ $\rightarrow$ $0.682\ K_2O$
$0.318\ CO_2$ **(melting loss)**
$1K_2O$ $\rightarrow$ $1.47K_2CO_3$
$1K_2CO_3$ $\rightarrow$ $1.2K_2CO_3.1.5H_2O$

POTASSIUM CARBONATE, HYDRATED

$K_2CO_3.1.5H_2O$

Hydrated potash. Pearl ash.
French: Carbonate de potassium hydraté
German: Hydratisierte Pottasche, Hydrat-Pottasche

White crystalline granular material. Less hygroscopic than the calcined salt.

QUALITY REQUIREMENTS
Na_2CO_3	max 0.6%
KCl	max 0.2%
K_2SO_4	max 0.1%
Fe_2O_3	max. 0.0003% (colourless glass)
Cr_2O_3	max. 0.00005% (colourless glass)

APPLICATION
Main raw material for K_2O in glass.

NOTE
Sometimes $K_2CO_3.2H_2O$, containing 54.06% K_2O has been used.

CONVERSION FACTORS:
$1K_2CO_3.1.5H_2O$	$\rightarrow$	$0.57K_2O$
		$0.43CO_2+H_2O$ (melting loss)
$1K_2O$	$\rightarrow$	$1.75K_2CO_3.1.5H_2O$

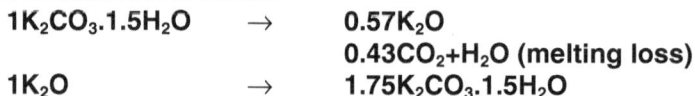

POTASSIUM CHLORIDE

KCl
French:　Chlorure de potassium
German:　Kaliumchlorid

White salt.

APPLICATION
Opalising agent in alabaster glass.

NOTE
Also potassium sulphate and alum are used for the same application.

POTASSIUM CHROMATE

K_2CrO_4
French:　Chromate de potassium
German:　Kaliumchromat

Yellow crystalline powder.

APPLICATION
Colouring agent for green glasses. (See also under 'Chromium oxide')

CONVERSION FACTORS
$1K_2CrO_4$　$\rightarrow$　$0.391Cr_2O_3$
　　　　　　　　　$0.485K_2O$
$1Cr_2O_3$　$\rightarrow$　$2.554K_2CrO_4$

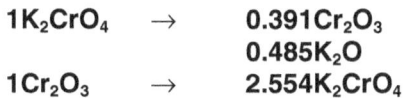

POTASSIUM HYDROXIDE

KOH
French: Hydroxyd de potassium
German: Kaliumydroxyd, Kalilauge

Strong attack on skin. Protect the eyes!

APPLICATION
Used as water solution, e.g. to prevent dusting in lead crystal batches where it is also said to speed up the melting rate.

CONVERSION FACTORS
     1KOH $\rightarrow$ $0.84K_2O$

POTASSIUM NITRATE

KNO_3

'Nitre'
French: Nitrate de potassium, salpêtre
German: Kaliumnitrat, Kalisalpeter

White crystalline powder. Potassium nitrate in contrast to sodium nitrate is not hygroscopic but tends to develop hard lumps during storage, especially when exposed to pressure.

QUALITY REQUIREMENTS
Pure nitre: min. 99.7% KNO_3, max. 0.001 Fe_2O_3.

APPLICATION
Refining and decolourising (oxidising) agent usually used in combination with arsenic trioxide, antimony trioxide or sodium sulphate.

NOTE
Sometimes calcium nitrate, $Ca(NO_3)_2$, or barium nitrate, $Ba(NO_3)_2$, have been used as refining agents, for instance in low alkali glasses. More common is sodium nitrate $NaNO_3$.

CONVERSION FACTORS

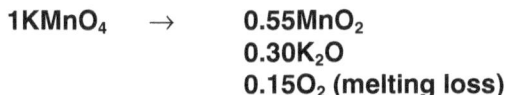

$1KNO_3 \rightarrow 0.466K_2O$
$0.534N_2O_5$ (melting loss)
$1K_2O \rightarrow 2.15KNO_3$

POTASSIUM PERMANGANATE

$KMnO_4$
French: Permanganate de potassium
German: Kaliumpermanganat

Dark violet crystalline powder, soluble in water. Gives off oxygen at about 240°C.

APPLICATION
Decolourising agent mostly used in pot melting. It has the advantage, compared with the MnO_2 mineral pyrolusite, of having a constant composition and very low iron content. Colouring agent for violet glasses or, in combination with Fe_2O_3, brown glasses.

CONVERSION FACTORS

$1KMnO_4 \rightarrow 0.55MnO_2$
$0.30K_2O$
$0.15O_2$ (melting loss)

PRASEODYMIUM OXIDE

Pr_6O_{11}
French: Oxyde de praseodyme
German: Praseodymoxyd

Black powder.

APPLICATION
Produces green-yellow colour in glass and shows 'dichroism'. In combination with Nd_2O_3; grey–olive-green–brown-red colours are obtainable.

SELENIUM

Se
French: Sélénium
German: Selen

Black powder.

APPLICATION
Main decolourising agent for soda-lime glasses when melted in tanks, sometimes in combination with small amounts of cobalt oxide. Colouring agent: 'selenium rose', 'selenium ruby' (in combination with cadmium sulphide).
See also: barium, sodium and zinc selenite.

NOTE
'Red selenium' is only another modification of the black selenium.

SILICON

Si
French: Silicium
German: Kiesel

Grey-black powder with metallic lustre.

APPLICATION
Silicon is used as a strong reducing agent for instance in some coloured glasses.

SILICON DIOXIDE

SiO_2

Silica sand
French: Acide silicique, sable
German: Kieselsäure, Sand

Silica sand consists essentially of quartz grains either from sand deposits (in some cases requiring chemical purification, washing and screening) or from crushed quartz.

QUALITY REQUIREMENTS
There are nearly always other mineral grains besides quartz in sands. Feldspar grains contribute to the alumina content of the glass but normally create no melting problems. Grains or heavy minerals containing iron, titanium or chromium can cause colour when colourless glasses are being produced. Often refractory alumina-containing minerals such as 'kyanite' appear in the sand and may result in small white stones especially in relatively less aggressive melts such as high lead glasses. It is recommended for lead crystal production to test the occurrence of such minerals by quick laboratory melts in platinum crucibles and counting the number of stones in a poured sample.

The sand ought to be free of clay minerals and similar impurities. Organic matter should be sieved away and magnetic particles should be separated by the use of a magnetic separator after the drying of the sand.

Normally the sand contains 3–5% moisture when delivered. If wet sand is used the water content must be determined continuously.

The main part of the sand grains should be between 0.2–0.4 mm. Grains >0.4 mm max. 2% and 0%>0.6 mm. <0.1 mm max. 1%.

ANALYTICAL REQUIREMENTS

Quality	Min. SiO_2 (%)	Max. Fe_2O_3 (%)	Max. Cr_2O_3 (%)
I	99.5	0.008	0.0001 (optical glass, heavy lead crystal)
II	99.0	0.014	0.0002 (ordinary 'colourless' glasses)
III	98.5	0.030	0.0003 (flat glass, bottles)
IV	According to agreement, analyses should be known.		

APPLICATION

Silica (SiO_2) is the main component in most commercial glasses and silica sand is the main raw material for this constituent.

SILVER NITRATE

$AgNO_3$
French: Nitrate d'argent
German: Silbernitrat

White salt, which will attack and cause a black stain on the skin.

APPLICATION

Used in 'silver-yellow' glasses. Added to the batch as water solution.

NOTE

Silver oxide Ag_2O, silver sulphide, Ag_2S or silver chloride $AgCl$ are often used in mixtures for so called silver staining. Silver halides are also used in photochromic glasses.

SODIUM ALUMINIUM FLUORIDE

Na_3AlF_6

Cryolite
French: Cryolite
German: Kryolith

Either the natural mineral or synthetic fluoride is used. Natural cryolite contains approx. 95% Na_3AlF_6. Bulk density ~1.2. Fe_2O_3~0.1%. Synthetic cryolite contains approx 99% Na_3AlF_6, 53.5% F and 0.05% Fe_2O_3. Bulk density ~0.7. It is important that synthetic cryolite really does contain the compound Na_3AlF_6 and not just a mixture of sodium and aluminium fluorides. Sodium fluoride is soluble in water, Na_3AlF_6 is not.

CONVERSION FACTORS

$1Na_3AlF_6 \rightarrow 0.544F_2$

$\rightarrow 0.440Na_2O$

$\rightarrow 0.240Al_2O_3$

APPLICATION

Main raw material in fluoride opal glass. Small quantities are sometimes used as fluxing agent (refining). During melting some of the fluorine is lost as SiF_6.

SODIUM ALUMINIUM SILICATE

$Na_2O.Al_2O_3.2SiO_2$

'Nepheline'
French: Néphéline
German: Nephelin

Grey or grey-green mineral powder. The composition often differs significantly from the theoretical. Here one example is given:
Nepheline: 43.2% SiO_2, 33.1% Al_2O_3, 1.3% Fe_2O_3, 15.7% Na_2O, 5.5% K_2O, 1.2% CaO

APPLICATION

Introduction of Al_2O_3 and alkali oxides. (See also under potassium–aluminium–silicate and sodium–potassium–aluminium–silicate.)

SODIUM CARBONATE, SODA ASH

Na_2CO_3
French: Carbonate de sodium (soude); 'soude'
German: Natriumkarbonat, wasserfrei; Kalzinierte Soda

White granulated grains or powder.
Hygroscopic. Water content ought to be checked continuously. (Dries at 250°C.)

QUALITY REQUIREMENTS
The most commonly used quality is 'dense granular soda ash' which should contain min. 99% Na_2CO_3 (min. 58% Na_2O). Bulk weight ~1. This quality is not so dusty as other qualities. 'Light soda ash' finds occasional use. It is normally cheaper but its use involves serious dusting problems when handled and melted (nasal irritation, superstructure corrosion, etc.). In connection with batch pelletising, light soda ash may have some advantages.

QUALITY RECOMMENDATIONS
Grain size

max 1%	>	1.5 mm
max 10%	>	1.0 mm
min 50%		0.2–0.6 mm
max 5%	<	0.1 mm

Composition

Total Na_2O content	>	58% Na_2O (99% Na_2CO_3)
Sodium chloride		max 0.5%
Sodium sulphate		max 0.1%
Iron as Fe_2O_3		max 0.005%
Insoluble in water		max 0.1%
Water		max 1%

The material must not contain colouring metal compounds.

APPLICATION
Main raw material for Na_2O in glass.

NOTE
Crystallised sodium carbonate $Na_2CO_3.10H_2O$ is not used in glass making.

CONVERSION FACTORS

$1Na_2CO_3$	→	**0.585Na_2O**
		0.415H_2O (melting loss)
$1Na_2O$	→	**1.71Na_2CO_3**

SODIUM CHLORIDE

NaCl
French: **Chlorure de soude**
German: **Natriumchlorid**

White crystalline salt.

APPLICATION
Refining agent for instance, in glasses which have to be melted under reducing conditions (cadmium yellow glasses, selenium ruby, etc.). Some NaCl leaves the melt and produces HCl in the waste gases which may cause corrosion problems.

SODIUM DICHROMATE

$Na_2Cr_2O_7.2H_2O$
French: **Bichromate de sodium**
German: **Natriumbichromat**

Red-green crystalline powder.

APPLICATION
Colouring agent for green glasses. See also potassium, dichromate and chromic oxide.

CONVERSION FACTORS
$1Na_2Cr_2O_7.2H_2O$ $\rightarrow$ $0.510Cr_2O_3$
$0.208Na_2O$
$0.282H_2O+O_2$ (melting loss)

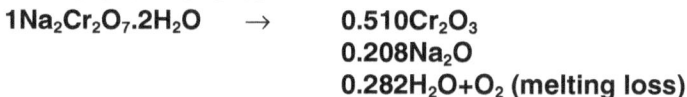

SODIUM DIURANATE

$Na_2U_2O_7$
French: Pyro-uranate de sodium
German: Natriumdiuranat, Uransaures Natrium

Orange-yellow powder.

APPLICATION
Introduction of U_3O_8 in glass. Colouring agent: yellow-green with green fluorescence.

NOTE
Sodium salts containing water are sometimes used.
$Na_2U_2O_7.3H_2O$ with 80% U_3O_8
$Na_2O_2O_7.6H_2O$ with 72% U_3O_8

CONVERSION FACTORS:
$$1Na_2U_2O_7 \rightarrow \quad 0.885U_3O_8$$
$$0.099Na_2O$$

SODIUM HYDROXIDE

NaOH
French: Hydrate de soude
German: Natriumhydroxid, Ätznatron

Sodium hydroxide is normally used as approx. 50% water solution. It will strongly attack skin and clothes. Very dangerous to the eyes! Hygroscopic.

APPLICATION
Used to replace some of the soda ash. Acts as binder for pelletising of soda-lime batches.

CONVERSION FACTORS
$$1NaOH \rightarrow 0.775Na_2O$$
$$1Na_2O \rightarrow 1.29NaOH$$

SODIUM METAPHOSPHATE

$NaPO_3$
French: Metaphosphate de sodium
German: Natriummetaphosphat

Glassy grains or flakes.

APPLICATION
Used as opalising agent in some phosphate opal glasses or to introduce P_2O_5 in special glasses.

CONVERSION FACTORS
$$1NaPO_3 \quad \rightarrow \quad 0.696P_2O_5$$
$$0.304Na_2O$$

SODIUM NITRATE

$NaNO_3$
French: Nitrate de sodium (soude), Salpetre de soude
German: Natriumnitrat, Natronsalpeter

White crystalline powder or grains. Hygroscopic. Sodium nitrate is sometimes called 'Chile saltpetre' but that is the name of the impure salt found in nature.

QUALITY REQUIREMENT
~99.5% $NaNO_3$

APPLICATION
Oxidising agent. Used for decolourising and refining in combination with arsenic or antimony trioxide. It is not used as main Na_2O raw material because of trouble with furnace corrosion and pollution.

CONVERSION FACTORS
$$1NaNO_3 \quad \rightarrow \quad 0.365Na_2O$$
$$\rightarrow \quad 0.635N_2O_5 \text{ (melting loss)}$$
$$1Na_2O \quad \rightarrow \quad 2.74NaNO_3$$

SODIUM PHOSPHATE

$Na_2HPO_4.2H_2O$
French:　Phosphate disodique
German:　Dinatriumphosphat

White salt. Hygroscopic. Takes up water from the atmosphere and changes to $Na_2HPO_4.12H_2O$ (19.8P_2O_5, 17.3Na_2O).

QUALITY REQUIREMENT
Commercial product contains about 40% P_2O_5.

APPLICATION
Opalising agent in phosphate opal glasses.

CONVERSION FACTORS

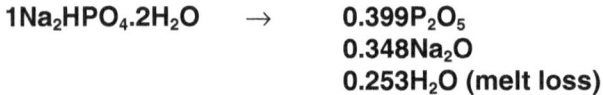

$1Na_2HPO_4.2H_2O$　　$\rightarrow$　　　0.399P_2O_5
　　　　　　　　　　　　　　　　　　　0.348Na_2O
　　　　　　　　　　　　　　　　　　　0.253H_2O (melt loss)

SODIUM POTASSIUM ALUMINIUM SILICATE

Nepheline syenite

Phonolite

(see also: Potassium aluminium silicate)
French: Syénite néphélinique, syénite à néphéline
German: Nephelinsyenit or Phonolit

Pulverised minerals of various composition; for instance

	German Phonolite (%)	Canadian nepheline syenite (%)	Norwegian nepheline syenite (%)
SiO_2	56.6	49.9	56.5
Al_2O_3	21.1	30.5	24.8
Fe_2O_3	2.8	0.3	0.1
K_2O	7.5	3.6	9.2
Na_2O	7.2	13.4	7.2
CaO	1.6	1.3	0.8
MgO	0.4	0.1	0.0
Loss of ignition	1.8	0.8	0.7

APPLICATION
Raw material for Al_2O_3 mostly used in bottle glasses and glass fibres. As can be seen from the analysis, some other components are introduced at the same time, for instance alkali oxides.

SODIUM SELENITE

Na_2SeO_3
French: Sélenite de sodium
German: Natriumselenit

White water soluble and hygroscopic salt. It tends to become rose-coloured by storage because of decomposition (Se). Toxic. Attack on skin by direct contact.

QUALITY REQUIREMENTS
99% Na_2SeO_3 (~45% Se, 1% H_2O)

APPLICATION
Decolourising and colouring agent (selenium rose, selenium ruby).

CONVERSION FACTORS
$1Na_2SeO_3$ $\rightarrow$ $0.358Na_2O$
                   $0.457Se$
1Se $\rightarrow$ $2.19Na_2SeO_3$

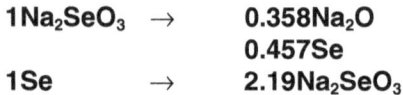

SODIUM SILICOFLUORIDE

Na_2SiF_6
French: Fluosilicate de sodium, silicofluorure de sodium
German: Kiselfluornatrium, Natriumhexafluorosilikat

White powder. Commercial quality 98% Na_2SiF_6.

APPLICATION

Opalising agent in fluoride opal glass. The fluorine losses by melting are higher than if for instance cryolite is used.

NOTE
When fluoride compounds are introduced in the batch the silica content of the glass is somewhat reduced by loss of SiF_6 from the melt.

CONVERSION FACTORS
$1Na_2SiF_6$ $\rightarrow$ $0.33Na_2O$
         $\rightarrow$ $0.32SiO_2$
         $\rightarrow$ $0.61F_2$

SODIUM SULPHATE, ANHYDROUS

Na_2SO_4
French: Sulfate de sodium
German: Natriumsulfat, 'Sulfat'

Hygroscopic white salt.

QUALITY REQUIREMENTS
'Pure' sulphate: min. 99.5% Na_2SO_4, 0.005% Fe_2O_3. For ordinary glass: min. 93% Na_2SO_4, max. 0.15% Fe_2O_3.

APPLICATION
The common refining agent for tank melting. Large amounts of sulphate do not dissolve in the glass melt if not mixed with a reducing agent such as carbon. Excess sulphate floats on the surface of the melt and is called 'glass gall'.

NOTE
Glauber's salt is $Na_2SO_4.10H_2O$ and is usually not used in glass melting.

CONVERSION FACTORS

$1Na_2SO_4$	$\rightarrow$	$0.436Na_2O$
		$0.564SO_3$
$1Na_2O$	$\rightarrow$	$2.29Na_2SO_4$

SODIUM SULPHIDE

$Na_2S.3H_2O$
French: Sulphide de sodium
German: Natriumsulfid, Schwefelnatron

Very hygroscopic, alkaline, bad smelling (H_2S) salt. Attack on skin.

QUALITY REQUIREMENTS
~60% $Na_2S.3H_2O$. Used in the form of flakes.

APPLICATION
Used in some sulphide coloured glasses (for instance yellow/amber).

STRONTIUM CARBONATE

$SrCO_3$
French: Carbonate de strontium
German: Strontiumkarbonat

White powder.

APPLICATION
Main raw material for SrO in glass, e.g. television tubes.

CONVERSION FACTORS
$1SrCO_3$	$\rightarrow$	$0.702SrO$
		$0.298CO_2$ (melting loss)
$1SrO$	$\rightarrow$	$1.425SrCO_3$

SULPHUR

S
French: Soufre
German: Schwefel

Yellow combustible powder.

APPLICATION
Used in some sulphide coloured glasses.

TIN OXIDE

SnO_2
French: Oxyde d'etain
German: Zinnoxyd, Zinndioxyd, Zinnasche

Light white powder.

APPLICATION
Tin oxide in small quantities is used in gold ruby glasses.
In copper ruby SnO, metallic tin powder or tin chloride $SnCl_2.2H_2O$ act as reducing agents.

TITANIUM DIOXIDE

TiO_2
French: Oxyde de titane
German: Titanoxyd

White powder or small granules. Contains usually ~98% TiO_2.

APPLICATION
Introduction of TiO_2 in glass. TiO_2 brings high refractive index (reflecting beads) and acts as a fluxing component (decreases viscosity). TiO_2 in combination with cerium oxide (CeO_2) is used to obtain cerium-titanium

yellow glasses. TiO_2 gives a light yellow tint also in glasses with very low iron content and can then cause decolourising problems. TiO_2 is used in some kinds of glass ceramics as a nucleation agent.

NOTE
'Rutile' is a mineral of which the main composition is titanium dioxide (70–90% TiO_2) but does also often contain iron, aluminium etc. The oxide is often used as a pigment and then sometimes mixed with other materials (Al, Zn, Si).

TRICALCIUM PHOSPHATE

$Ca_3(PO_4)_2$
French: Phosphate tricalcique
German: Trikalziumphosphat

White powder. Commercial products often contain lower P_2O_5% than the theoretical figure of the chemical formula. Actual analysis should be obtained from supplier.

QUALITY REQUIREMENTS
Synthetic calcium phosphate with low iron content or a material resulting from the burning of bones, 'bone ash', are used in the glass industry. Bone ash usually contains some calcium fluoride and organic material. Examples: ~85% $Ca_3(PO_4)_2$, ~12% $CaCO_3$, ~2% CaF_2 (calcium fluoride), ~1% $MgCO_3$.

APPLICATION
Opalising agent in phosphate opal glasses. Introduction of P_2O_5 (and CaO) in glass.

CONVERSION FACTORS
$1Ca_3(PO_4)_2 \rightarrow$ **0.56CaO (at 85% purity and including CaO from carbonate and fluoride)**
    $0.39P_2O_5$

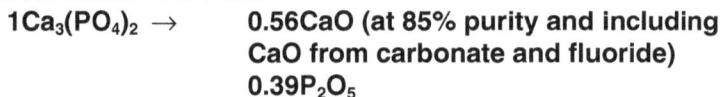

TUNGSTEN OXIDE

WO_3
French: Oxyde de tungsténe
German: Wolframoxyd, Wolframsäure

Yellow or orange-yellow powder.

APPLICATION
Used as melting accelerator, e.g. in low alkali borosilicate glasses. Influences devitrification properties (nucleation) and is sometimes used in selenium-ruby glasses.

NOTE
Sodium tungstate $Na_2WO_4.2H_2O$(18.78% Na_2O, 70.3% WO_3) has been used for the same purposes.

URANIUM OXIDE

U_3O_8
French: Oxyde d'uranium
German: Uranyluranat

Yellow-brown powder.

APPLICATION
Colouring agent. Produces green-yellow colour in glass (green fluorescence).

NOTE
U_3O_8 is sometimes exchanged for UO_3 or UO_2 (black powder).

VANADIUM PENTOXIDE

V_2O_5
French: Oxyde de vanadium
German: Vanadinpentoxyde

Yellow red-brown powder.

APPLICATION
Introduction of vanadium oxide in glass (e.g. UV absorbing glasses). Gives a green colour in glass similar to Cr_2O_3-green but the colouring effect is much lower.

NOTE
V_2O_5 reduces surface tension in the melt. Acts as catalyser on SO_2 resulting in corrosive sulphuric acid in waste gases.

ZINC OXIDE

ZnO
French: Oxyde de zinc
German: Zinkoxyd

White powder. Bulk density ~0.8.

QUALITY REQUIREMENTS
Zinc oxide does also often contain some lead oxide (PbO). This may cause trouble, e.g. in selenium-ruby batches or problems at the electrodes in electric melting. For selenium-ruby a maximum of 0.4% PbO can be accepted. Commercial ZnO does often contain some colouring metal oxides (CoO, CuO, etc.) which cause discolouration in the production of colourless glasses. Actual analysis should be obtained from supplier.

APPLICATION
Introduction of zinc oxide (ZnO) in glass. Increases chemical durability (e.g. resistance in dish washing machines) and decreases the melt viscosity. Component in fluoride-opal glasses, selenium-ruby and other sulphide coloured glasses, optical glasses and solder glasses.

ZINC SELENITE

$ZnSeO_3$
French: Sélenite de zinc
German: Zinkselenit

White or yellow-white powder. Commercial products contain 40–41% Se.

APPLICATION
Decolourising agent. Easy to handle and is said to give less selenium loss than metallic Se in melting.

CONVERSION FACTORS
$1ZnSeO_3$ $\rightarrow$ **0.411Se**
                    $\rightarrow$ **0.423ZnO**
                    $\rightarrow$ **0.166O_2 (melting loss)**
1Se $\rightarrow$ **2.43 $ZnSeO_3$**

ZIRCON SILICATE, ZIRCON SAND

$ZrO_2.SiO_2$
French: Zircon, silicate de zirconium
German: Zirkonsilikat, 'Zirkonit'

Natural mineral: 'Zircon sand'. Composition varies. Example: 62–63% ZrO_2, 36–37% SiO_2, max. 0.3% Al_2O_3, max. 0.15% Fe_2O_3, max. 0.3% TiO_2. A fine powdered product is recommended to increase solution in the glass melt.

APPLICATION
Introduction of ZrO_2 in glass. Increase refractive index and chemical durability (e.g. against alkaline solutions).

NOTE
Various zirconates or even the oxide ZrO_2 are sometimes used.

CONVERSION FACTORS
$1ZrO_2.SiO_2$ $\rightarrow$ **0.672ZrO_2**
                        $\rightarrow$ **0.328SiO_2**
$1ZrO_2$ $\rightarrow$ **1.49$ZrO_2.SiO_2$**

CONVERSION FACTORS FOR SOME COMMON GLASS RAW MATERIALS

I   Theoretical figures, 100% pure materials
II   Normal approx. figures in commercial materials. If no figures are given here, the figures in 'I' can be used
*   Composition varies. Use figures from analysis of actual material

Material	% glass component				kg material to get 1 kg of component in glass	
	I		II			
Aluminium hydrate Al(OH)₃	65.4	Al₂O₃	65	Al₂O₃	1.53	Al₂O₃
Antimony oxide Sb₂O₃	100	Sb₂O₃	99	Sb₂O₃	1.00	Sb₂O₃
Arsenic As₂O₃	100	As₂O₃	99.5	As₂O₃	1.00	As₂O₃
Barium carbonate BaCO₃	77.7	BaO	77.0	BaO	1.29	BaO
Barium selenite BaSeO₃	29.9	Se	28.5	Se		
	58.0	BaO		BaO		
Borax anhydrous Na₂B₄O₇	69.2	B₂O₃	69	B₂O₃	1.45	B₂O₃
	30.8	Na₂O	30.8	Na₂O		
Borax, cryst Na₂B₄O₂.10H₂O	36.5	B₂O₃	36.5	B₂O₃	2.73	B₂O₃
	16.3	Na₂O	16.3	Na₂O		
Boric acid H₃BO₃	56.3	B₂O₃	56.3	B₂O₃	1.79	B₂O₃

Material				Factor	Oxide
Dolomite MgCO$_3$.CaCO$_3$	21.7	20*	MgO	4.61	MgO
	30.4	31	CaO		
Fluorspar CaF$_2$	71.8		CaO	2.05	F$_2$
	48.7	47*	F$_2$		
Feldspar (Pegmanite Forshammar)		75.9*	SiO$_2$	7	Al$_2$O$_3$
		14.3	Al$_2$O$_3$		
		4.8	K$_2$O		
		4.0	Na$_2$O		
Limestone CaCO$_3$	56.0	55*	CaO	1.78	CaO
Cryolite (nat)		54.3*	F$_2$	1.84	F$_2$
		44.3	Na$_2$O		
		24.3	Al$_2$O$_3$		
Cryolite (synth.) Na$_3$AlF$_6$	54.2		F$_2$	1.84	F$_2$
	44.2		Na$_2$O		
	24.1		Al$_2$O$_3$		
Calcium borate (Colemanite)		47*	B$_2$O$_3$	2.13	B$_2$O$_3$
		26.5*	CaO		
		2	MgO		
		3	SiO$_2$		
Lithium carbonate Li$_2$CO$_3$	40.4		Li$_2$O	2.47	Li$_2$O
Red lead Pb$_3$O$_4$	97.7		PbO	1.02	PbO
Lead monosilicate PbO.SiO$_2$	85		PbO	1.18	PbO
	15		SiO$_2$		

Material	% glass component			kg material to get 1 kg of component in glass	
	I	II			
Litharge PbO	100		PbO	1.0	PbO
Potash (calc.) K_2CO_3	68.2		K_2O	1.47	K_2O
Potash (hydr.) 'Pearl ash' $K_2CO_3 \cdot 1.5H_2O$	57.0		K_2O	1.75	K_2O
Potassium bicarbonate $KHCO_3$	47.1		K_2O	2.12	K_2O
Potassium nitrate KNO_3	46.5		K_2O	2.15	K_2O
Potassium bichromate $K_2Cr_2O_7$	51.7	32.0	Cr_2O_3 / K_2O	1.93	Cr_2O_3
Potassium permanganate $KMnO_4$	55.0	30.0	MnO_2 / K_2O	1.82	MnO_2
Copper sulphate $CuSO_4 \cdot 5H_2O$	31.8		CuO	3.14	CuO
Rasorite (nat.) Sodium borate		47* / 21.5 / 0.4 / 15.5 / 0.3	B_2O_3 / CaO / Al_2O_3 / Na_2O / Al_2O_3	2.13	B_2O_3
Sodium sulphate Na_2SO_4	43.6		Na_2O	2.3	Na_2O

Material	%	Component	%	Component	Factor	Component
Sodium selenite Na_2SeO_3	45.7	Se	35.8	Na_2O	2.8	Se
Sodium carbonate Na_2CO_3	58.5	Na_2O			1.71	Na_2O
Sodium carbonate 98%	57.4	Na_2O			1.74	Na_2O
Sodium carbonate 96%	56.1	Na_2O			1.78	Na_2O
Sodium nitrate $NaNO_3$	36.5	Na_2O			2.74	Na_2O
Zinc oxide	100	ZnO			1.00	ZnO
Zinc selenite	40.5	Se			2.43	Se
$ZnSeO_3$	41.1	ZnO	42.3	ZnO		

THE CALCULATION OF BATCH AND GLASS

COMPOSITIONS

By using the conversion factors given, the amount of each oxide in the glass can be calculated from the batch composition. The percentage glass composition is found by dividing the weight of each oxide from the raw materials by the weight of all the oxides and multiplying by 100.

To calculate a batch composition from a given percentage glass composition requires some experience in glass technology, as it is necessary to select materials which give favourable melting and refining conditions. For instance some of the sodium oxide may be introduced as soda ash and some as sodium nitrate or sulphate. In practice such factors as carry-over, volatilisation from the melt (PbO, F_2) and solution of refractory material (Al_2O_3, ZrO_2) must also be taken into account based on experiences in actual melting conditions.

If a material contributes more than one oxide to the glass, the calculation should start with this oxide. If, for instance, not enough B_2O_3 can be introduced as borax to get the required Na_2O percentage, then the rest of the B_2O_3 can be introduced as boric acid.

The table of conversion factors gives theoretical figures for 100% pure materials. As shown in the example below, using these in calculations, yield only approximate working values. Normal commercial figures are also given in the table but for exact calculations the actual chemical analysis is required.

Example 1. Calculation of the percentage glass composition from the batch for a heavy lead crystal:

100 kg dry sand	$100 \times 0.99 = 99.0 SiO_2$
60 kg red lead	$60 \times 0.977 = 58.6\ PbO$
40 kg hydrated potassium carbonate	$40 \times 0.57\ \ = 22.8 K_2O$
4 kg potassium nitrate	$4 \times 0.465\ = 1.9\ K_2O$
1 kg arsenic (As_2O_3)	$1 \times 1.16\ \ \ = \underline{\ \ \ 1.1 As_2O_5\ \ }$
	183.4 kg of glass

Theoretical composition:

$\dfrac{100 \times 99}{183.4} = {\sim}54.0\%\ SiO_2$ $\dfrac{100 \times 58.6}{183.4} = {\sim}32.0\%\ PbO$

$\dfrac{100 \times 24.7}{183.4} = {\sim}13.5\%\ K_2O$ $\dfrac{100 \times 1.1}{183.4} = {\sim}0.6\%\ As_2O_5$

The real composition will vary somewhat from the theoretically calculated values. First of all because of the volatilisation of lead (approximately 2–5% of the PbO content) and then because of the introduction of the products of corrosion from the pots (approximately 0.1–0.2% Al_2O_3 as well as a corresponding quantity of Fe_2O_3). An analysis of the glass would therefore give the following result

54.6% SiO_2	31% PbO	13.7% K_2O	0.2% Al_2O_3
0.5% As_2O_5	0.015% Fe_2O_3		

Example 2. Calculation of batch from a soda lime silica glass composition

73.1% SiO_2	1.3% Al_2O_3	7.5% CaO
1.4% MgO	15.8% Na_2O	0.9% K_2O

Feldspar will be used for the introduction of alumina with the analysis given in the conversion table. MgO is introduced with dolomite with an analysis of 30.4% CaO and 21.7% MgO (ignition loss 49.9%). When using raw materials which give more than one component (in this case feldspar or dolomite) the calculations should be carried out as follows.
1.3 kg of Al_2O_3 is obtained from 7×1.3=9.1 kg of feldspar
One obtains from the feldspar at the same time:

$$\frac{9.1 \times 75.9}{100} = 6.91 \text{ kg } SiO_2$$

$$\frac{9.1 \times 4.8}{100} = 0.44 \text{ kg } K_2O$$

$$\frac{9.1 \times 4.0}{100} = 0.36 \text{ kg } Na_2O$$

1.4 kg MgO is obtained from 1.4×4.61=6.45 kg dolomite.
One obtains from the dolomite at the same time:

$$\frac{6.45 \times 30.4}{100} = 1.96 \text{ kg CaO}$$

We also have to introduce 73.1–6.91=66.19 kg SiO_2 in the form of sand which is presumed to hold 99% SiO_2 so:

$$\frac{66.19 \times 100}{99} = 66.9 \text{ kg sand}$$

7.5–1.96 = 5.54 kg CaO in the form of lime, calcium carbonate, which is presumed to hold 55% CaO so

$$\frac{5.54 \times 100}{55} = 10.1 \text{ kg lime}$$

15.8–0.36 = 15.44 kg Na_2O in the form of soda which is presumed to hold 98% carbonate i.e. 57.4% Na_2O so:

$$\frac{15.44 \times 100}{57.4} = 26.9 \text{ kg soda}$$

0.9–0.44 = 0.46 kg K_2O in the form of potassium nitrate which is presumed to hold 46.5% K_2O so:

$$\frac{0.46 \times 100}{46.5} = 1 \text{ kg potassium nitrate}$$

The batch now looks like this:

		/100 kg sand	Rounded off
Sand	66.9 kg	100.0 kg	100.0 kg
Feldspar	9.1 kg	13.6 kg	13.5 kg
Dolomite	6.5 kg	9.7 kg	9.5 kg
Lime	10.1 kg	15.1 kg	15.0 kg
Soda	26.9 kg	40.2 kg	40.0 kg
Potassium nitrate	1.0 kg	1.0 kg	1.0 kg

COLOURED GLASSES

Examples of colour combinations as a guidance to the manufacture of coloured glasses. (Different possibilities are given for the same colour type.)

Blue	Pure blue	$CuO+CoO$
	Blue with a violet tint	CoO
	Sky blue	CuO
	Blue green	Fe_2O_3+CoO
		$CuO+Cr_2O_3$
		FeO
Brown	Red brown to yellow brown	$MnO_2+Fe_2O_3$
	Carbon amber	$C+Na_2SO_4(+FeS_2)$
	Brown orange	Se (red.)$+Fe_2O_3$
Grey	Grey	NiO (in soda glass)$+MnO_2+Fe_2O_3+CuO$
Green	Antique green	$Fe_2O_3(+C+Cr_2O_3)$
	Grass green	Cr_2O_3+CuO
	Yellow green	Cr_2O_3
		Pr_6O_{11}
		U_3O_8 (fluorescent green)
	Green blue	FeO
		Cr_2O_3+CuO
	Moss green	Cr_2O_3+CoO
		Cr_2O_3+NiO
	Olive green	$Fe_2O_3+Cr_2O_3$
		$Cr_2O_3+MnO_2$
Yellow	Yellow with green tint	CdS
	Intense yellow	$U_3O_8+Pr_6O_{11}$
	Gold yellow	$Se+MnO_2$ (in lead bearing glass)
	Canary yellow	$TiO_2+Ce_2O_3(3:2)$
	Amber yellow	MnO_2+FeO
		Se (in lead glass)
Orange	Orange red	$CdS+Se$
	Orange brown	MoS_2+S
		Se (red.)
Rose	Rose	Se (<u>not</u> in lead bearing glass)
Red	Ruby	$Se+CdS$ (<u>not</u> in lead bearing glass)
		Au
		Cu_2O (red.)
	Red violet	MnO_2+Se
	Wine red	Nd_2O_3+Se
Black	Black	FeS
		$MnO_2+Fe_2O_3+NiO+CoO+CuO$
		$Se+CoO$
Violet	Red violet	MnO_2 (in soda glass)
	Blue violet	MnO_2 (in potash glass)
	Violet	NiO (potash and lead glass)
	Rose violet	Nd_2O_3
		MnO_2+CoO

CHEMICAL ELEMENTS NAME, SYMBOL AND ATOMIC WEIGHT

Name	Symbol	Atomic weight	Name	Symbol	Atomic weight
Aluminium	Al	26.9	Mercury	Hg	200.6
Antimony	Sb	121.8	Molybdenum	Mo	95.9
Arsenic	As	74.9	Neodymium	Nd	144.2
Barium	Ba	137.4	Nickel	Ni	58.7
Beryllium	Be	9.01	Niobium	Nb	92.9
Bismuth	Bi	209	Nitrogen	N	14
Boron	B	10.9	Oxygen	O	16
Cadmium	Cd	112.4	Phosphorus	P	31
Calcium	Ca	40.1	Platinum	Pt	195.1
Carbon	C	12	Potassium	K	39.1
Cerium	Ce	140.1	Praseodymium	Pr	140.9
Caesium	Cs	132.9	Rhenium	Re	186.2
Chlorine	Cl	35.5	Rubidium	Rb	85.5
Chromium	Cr	52	Samarium	Sm	150.4
Cobalt	Co	58.9	Selenium	Se	78.9
Copper	Cu	63.5	Silicon	Si	28
Erbium	Er	167.3	Silver	Ag	107.9
Europium	Eu	152	Sodium	Na	23
Fluorine	F	19	Strontium	Sr	87.6
Gallium	Ga	69.7	Sulphur	S	32
Germanium	Ge	72.6	Tantalum	Ta	180.9
Gold	Au	197	Tellurium	Te	127.6
Hydrogen	H	1	Thallium	Tl	204.4
Indium	In	114.8	Thorium	Th	232
Iodine	I	126.9	Tin	Sn	118.7
Iridium	Ir	192.2	Titanium	Ti	47.9
Iron	Fe	55.8	Tungsten	W	183.9
Lanthanum	La	138.9	Uranium	U	238
Lead	Pb	207.2	Vanadium	V	50.9
Lithium	Li	6.9	Yttrium	Y	88.9
Magnesium	Mg	24.3	Zinc	Zn	65.4
Manganese	Mn	54.9	Zirconium	Zr	91.2

MELTING POINTS OF SOME GLASS MAKING COMPOUNDS

NAME	FORMULA	NOTE	°F	°C
Aluminium oxide	Al_2O_3		3720.0	2050.0
Aluminium silicate	$3Al_2O_3.2SiO_2$		3290.0	1810.0
Barium carbonate	$BaCO_3$	d.	2640.0	1450.0
Barium nitrate	$Ba(NO_3)_2$		1100.0	592.0
Barium silicate	$BaSiO_3$		2920.0	1605.0
Barium sulphate	$BaSO_4$		2875.0	1580.0
Boric acid	H_3BO_3	d.	365.0	185.0
Boron oxide crystal	B_2O_3	c.	910.0	490.0
Calcia	CaO		4660.0	2572.0
Calcium carbonate	$CaCO_3$	d.	Bright red heat	Bright red heat
Calcium fluoride	CaF_2		2480.0	1360.0
Calcium nitrate	$Ca(NO_3)_2$		1040.0	560.0
Calcium pyrophosphate	$Ca_2P_2O_7$		2245.0	1230.0
Tricalcium phosphate	$Ca_3(PO_4)_2$		3040.0	1670.0
Calcium silicate	$CaSiO_3$		2805.0	1540.0
Calcium sulphate	$CaSO_4$		2640.0	1450.0
Chromic oxide	Cr_2O_3		3615.0	1990.0
Copper (metal)	Cu		1981.0	1083.0
Cryolite	Na_3AlF_6		1830.0	1000.0
Ferrous silicate	$FeSiO_3$		2200.0	1200.0
Lead (metal)	Pb		620.0	327.0
Lead metaborate	$Pb(BO_2)_2$		Low red heat	Low red heat
Lead oxide (litharge)	PbO		1630.0	888.0
Lead oxide (red lead)	Pb_3O_4	d.	930.0	500.0
Lead silicate	$PbSiO_3$		1410.0	766.0
Lithium carbonate	Li_2CO_3		1145.0	618.0
Lithium nitrate	$LiNO_3$		490.0	255.0
Lithium silicate	$LiSiO_3$		2190.0	1200.0
Lithium sulphate	Li_2SO_4		1580.0	860.0
Magnesium carbonate	$MgCO_3$	d.	Low red heat	Low red heat
Magnesium oxide	MgO		5070.0	2800.0
Magnesium silicate	$MgSiO_3$		2835.0	1555.0
Magnesium sulphate	$MgSO_4$		2165.0	1185.0
Phosphorus pentoxide	P_2O_5		1045.0	563.0
Platinum (metal)	Pt		3227.0	1775.0

NAME	FORMULA	NOTE	°F	°C
Potassium carbonate	K_2CO_3		1635.0	891.0
Potassium hydroxide	KOH		716.0	380.0
Potassium nitrate	KNO_3		635.0	334.0
Potassium silicate	K_2SiO_3		1790.0	976.0
Potassium sulphate	K_2SO_4		1970.0	1076.0
Selenium	Se	Boils	1270.0	688.0
Selenium dioxide	SeO_2		644.0	340.0
Silica	SiO_2		3135.0	1725.0
Silver (metal)	Ag		1761.0	960.5
Sodium aluminate	$NaAlO_2$		3000.0	1650.0
Sodium diborate	$Na_2B_4O_7$		1365.0	741.0
Sodium carbonate	Na_2CO_3		1562.0	851.0
Sodium chloride	NaCl		1480.0	804.0
		Boils	2575.0	1413.0
Sodium nitrate	$NaNO_3$		587.0	308.0
Sodium silicate	Na_2SiO_3		1990.0	1089.0
Sodium disilicate	$Na_2Si_2O_5$		1605.0	874.0
Sodium sulphate	Na_2SO_4		1624.0	884.0
Sulphur	S		827.0	444.6
Zinc oxide	ZnO		u.p.>3270	u.p.>1800
Zinc silicate	$ZnSiO_3$		2620.0	1437.0
Zirconia	ZrO_2		4890.0	2700.0
Zirconium silicate (zircon)	$ZrSiO_4$		4620.0	2550.0

d   decomposes or loses oxygen
u.p. under pressure
c   approximate softening temperature of glassy form